BIBLIOGRAPHY

OF PAPERS

PUBLISHED BY L. ZECHMEISTER AND CO-AUTHORS
IN THE FIELDS OF CHEMISTRY AND BIOCHEMISTRY

1913–1958

SPRINGER-VERLAG WIEN GMBH
1958

ISBN 978-3-7091-2033-0 ISBN 978-3-7091-4207-3 (eBook)
DOI 10.1007/978-3-7091-4207-3

This Bibliography is dedicated to the following Co-authors:

M. Bálint (Hungary), J. Bársony (Hungary), R. Bender (Germany), T. Béres (Hungary), N. Bjerrum (Denmark), J. Bonner (U. S. A.), W. V. Bush (U. S. A.), C. E. Calbert (U. S. A.), A. Chatterjee (India), L. v. Cholnoky (Hungary), J. Csabay (Hungary), D. L. Currell (U. S. A.), J. Dale (Norway), the late H. J. Deuel, Jr. (U. S. A.), E. Ernst (Hungary), R. B. Escue (U. S. A.), H. L. Falk (U. S. A.), C. Faurholt (Denmark), P. Fischer Jörgensen (Denmark), D. L. Fox (U. S. A.), O. Frehden (Hungary), T. Fukui (Japan), P. Fürth (Hungary), J. Ganguly (India), Ch. Gansser (Switzerland), J. C. Gjaldbaek (Denmark), W. Grassmann (Germany), S. M. Greenberg (U. S. A.), A. J. Haagen-Smit (U. S. A.), F. T. Haxo (U. S. A.), C. Hendrick (U. S. A.), G. de Hevesy (Sweden), H. H. Inhoffen (Germany), B. v. Issekutz (Hungary), C. H. Johnston (U. S. A.), G. Karmakar (India), the late W. Kindler (Germany), J. S. Kittredge (U. S. A.), A. Koch (Hungary), B. K. Koe (U. S. A.), P. Kotin (U. S. A.), J. Leemann (Switzerland), R. M. Lemmon (U. S. A.), the late A. L. LeRosen (U. S. A.), W. Lijinsky (England), K. Lunde (Norway), W. H. McNeely (U. S. A.), E. F. Magoon (U. S. A.), H. Mark (U. S. A.), E. R. Meserve (U. S. A.), E. Neumann (Hungary), C. B. van Niel (U. S. A.), L. Pauling (U. S. A.), F. J. Petracek (U. S. A.), the late I. Pfeifer (Hungary), J. H. Pinckard (U. S. A.), I. Pinczési (Hungary), A. Polgár (U. S. A.), M. Rohdewald (Germany), P. Rom (Hungary), A. Sandoval (Mexico), W. A. Schroeder (U. S. A.), J. W. Sease (U. S. A.), M. B. Shimkin (U. S. A.), G. Sólyom (Hungary), R. Stadler (Germany), W. T. Stewart (U. S. A.), E. Straub (U. S. A.), E. Sumner (U. S. A.), P. Szécsi (Hungary), K. Szilárd (Hungary), Y. W. Tang (China), G. Tóth (Hungary), J. Truka (Hungary), K. Tsukida (Japan), P. Tuzson (Hungary), E. Ujhelyi (Hungary), A. Unmack (Denmark), É. Vajda (Hungary), V. Vrabély (Hungary), L. Wallcave (U. S. A.), F. W. Went (U. S. A.), B. Wille (Denmark), and the late R. Willstätter (Germany).

The co-operation of the Editors and Publishers of the following periodicals is appreciated:

Acta Chemica Scandinavica. American Scientist. Analytical Chemistry. Annals of the New York Academy of Sciences. Archives of Biochemistry and Biophysics. Berichte der deutschen chemischen Gesellschaft. Biochemical Preparations. Biochemische Zeitschrift. Boletin del Instituto de Química (Mexico). Bulletin de la Société de chimie biologique. Chemical and Engineering News. Chemical Reviews. Chemiker-Zeitung. Chemistry and Industry. Ciencia e investigación (Buenos Aires). Det Kgl. Danske Videnskabernes Selskab Mat.-Fys. Meddelelser. Die Naturwissenschaften. Discussions of the Faraday Society. Enzymologia. Ergebnisse der Physiologie, biologischen Chemie und experimentellen Pharmakologie. Experientia. Fortschritte der Chemie organischer Naturstoffe. Helvetica Chimica Acta. Hoppe-Seyler's Zeitschrift für physiologische Chemie. Isis (U. S. A.). Justus Liebigs Annalen der Chemie. Kagaku (Japan). Magyar Chemiai Folyóirat. Matematikai és Természettudományi Értesítő. Monatshefte für Chemie. Nature (London). Nordisk Handelsblad for kemisk Industri. Orvosi hetilap. Plant Physiology. Proceedings of the National Academy of Sciences (U. S. A.). Science (Washington). Sitzungsberichte der kgl. preußischen Akademie der Wissenschaften. The Biochemical Journal. The Harvey Lectures. The Journal of Biological Chemistry. The Journal of the American Chemical Society. Triangle (Basel). Vitamins and Hormones. Zeitschrift für Elektrochemie. Zeitschrift für physikalische Chemie. Zentralblatt für Mineralogie, Geologie und Paläontologie.

Most of our experiments were conducted in the following laboratories.

Chemisches Laboratorium der Eidgenössischen Technischen Hochschule, Zürich.

Kaiser-Wilhelm-Institut für Chemie, Berlin-Dahlem.

Royal Veterinary and Agricultural Academy, Copenhagen.

Chemical Institute, Medical School, The University of Pécs.

Gates and Crellin Laboratories of Chemistry, California Institute of Technology, Pasadena, California.

Contents.

Page

Carbohydrates 1
Enzymes 2
Anthocyanes 3
Carotenoids in General 3
Plant Carotenoids 4
Animal Carotenoids 7
Carotenoids: Structural Conversions in vitro 8
Carotenoids: Stereochemistry 9
Naturally Occurring *cis* and Poly*cis* Carotenoids 10
Provitamin A Potency and Structure 11
Provitamin A Potency and Steric Configuration 12
Naturally Occurring Colorless Polyenes 13
Miscellaneous Colorless Plant Products 14
Plant Physiology and Chemical Genetics 14
cis-trans Isomerization of Diphenylpolyenes, Phenyl-polyenaldehydes, Azines, and Cyanines 15
Carcinogens and Related Compounds 15
Light Aromatic Hydrocarbons 16
Organic Lead Compounds 16
Organic Minerals 16
Chromatographic Methods 17
Reduction Methods 18
Effect of Ultrasonic Waves on Organic Compounds 19
Electrolytic Dissociation 19
Radioactive Isotopes 19
Textbooks 20
Author Index 21

Carbohydrates.

1. L. ZECHMEISTER: Zur Kenntnis der Cellulose und des Lignins. Thesis. Zürich (1913).
2. R. WILLSTÄTTER und L. ZECHMEISTER: Zur Kenntnis der Hydrolyse von Cellulose. I. Ber. dtsch. chem. Ges. **46**, 2401–2412 (1913).
3. R. WILLSTÄTTER und L. ZECHMEISTER: Zur Kenntnis der Hydrolyse von Cellulose (II. Mitt.). Ber. dtsch. chem. Ges. **62**, 722–725 (1929).
4. L. ZECHMEISTER und G. TÓTH: Zur Kenntnis der Hydrolyse von Cellulose und der dabei auftretenden Zwischenprodukte. (III. Mitt. in der von R. WILLSTÄTTER und L. ZECHMEISTER begonnenen Reihe.) Ber. dtsch. chem. Ges. **64**, 854–870 (1931).
   Mat. termtud. ért. **48**, 443–458 (1931).
5. L. ZECHMEISTER und G. TÓTH: Partieller Abbau von tierischer Cellulose. (IV. Mitt. in der von R. WILLSTÄTTER und dem erstgenannten Verf. begonnenen Reihe.) Zeitschr. physiol. Chem. (Hoppe-Seyler) **215**, 267–276 (1933).
   Mat. termtud. ért. **50**, 430–440 (1933).
6. L. ZECHMEISTER, H. MARK und G. TÓTH: Cellotriose und ihre Bedeutung für das Strukturbild der Cellulose. Ber. dtsch. chem. Ges. **66**, 269–275 (1933).
7. L. ZECHMEISTER und G. TÓTH: Über Cellotriose (Bemerkungen zu den jüngsten Arbeiten der HHrn. K. Heß und K. Dziengel, sowie C. Trogus und K. Heß). Ber. dtsch. chem. Ges. **68**, 2134–2136 (1935).
8. L. ZECHMEISTER: Zur Einwirkung von Acetylbromid auf Cellulose. Ber. dtsch. chem. Ges. **56**, 573–578 (1923).
9. L. ZECHMEISTER und G. TÓTH: Über die Polyose der Hefemembran. I. Biochem. Z. **270**, 309–316 (1934).
   Mat. termtud. ért. **51**, 271–282 (1934).
10. L. ZECHMEISTER und G. TÓTH: Über die Polyose der Hefemembran. II. Biochem. Z. **284**, 133–138 (1936).
11. L. ZECHMEISTER: Zur Kenntnis des optischen Drehungsvermögens von Zuckerarten in Salzsäure. Z. physik. Chem. **103**, 316–336 (1922).
    Nordisk Handelsbl. kem. Ind. **4**, 39 (1923).
    Magy. chem. foly. **30**, 33–41 (1924).
12. L. ZECHMEISTER: Zur Einwirkung von konzentrierter Salzsäure auf Zuckerarten. Naturwiss. **15**, 73 (1927).

13. L. Zechmeister und G. Tóth: Chitin und seine Spaltprodukte. Fortschr. Chem. organ. Naturstoffe **2**, 212–247 (1939).
14. L. Zechmeister und G. Tóth: Zur Kenntnis der Hydrolyse von Chitin mit Salzsäure. (I. Mitt.) Ber. dtsch. chem. Ges. **64**, 2028–2032 (1931).
15. L. Zechmeister und G. Tóth: Zur Kenntnis der Hydrolyse von Chitin mit Salzsäure (II. Mitt.). Ber. dtsch. chem. Ges. **65**, 161–162 (1932).
Magy. chem. foly. **38**, 33–41 (1932).
16. L. Zechmeister und G. Tóth: Vergleich von pflanzlichem und tierischem Chitin. Z. physiol. Chem. (Hoppe-Seyler) **223**, 53–56 (1934). Mat. termtud. ért. **51**, 260–265 (1934).
17. G. Tóth and L. Zechmeister: Chitin Content of the Mandible of the Snail *(Helix pomatia)*. Nature (London) **144**, 1049 (1939).
18. L. Zechmeister und I. Pinczési: Octaacetyl-chitobiose aus Käfern. Z. physiol. Chem. (Hoppe-Seyler) **242**, 97–99 (1936).
19. L. Zechmeister und G. Tóth: Ein Beitrag zur Desamidierung des Glucosamins. Ber. dtsch. chem. Ges. **66**, 522–525 (1933).
Mat. termtud. ért. **49**, 134–141 (1932).
20. L. Zechmeister und G. Tóth: Zur Einwirkung von flüssigem NH_3 auf Zuckerderivate. Naturwiss. **23**, 35 (1935).
21. L. Zechmeister und G. Tóth: Einwirkung von flüssigem Ammoniak auf Octaacetyl-cellobiose. Liebigs Ann. Chem. **525**, 14–24 (1936).

Enzymes.

22. W. Grassmann, L. Zechmeister, G. Tóth und R. Stadler: Zur Spezifität polysaccharidspaltender Enzyme. Naturwiss. **20**, 639 (1932).
23. W. Grassmann, L. Zechmeister, G. Tóth und R. Stadler: Über den enzymatischen Abbau der Cellulose und ihrer Spaltprodukte. II. Mitt. über enzymatische Spaltung von Polysacchariden. Liebigs Ann. Chem. **503**, 167–179 (1933).
Magy. chem. foly. **39**, 59—64 (1933).
24. L. Zechmeister, G. Tóth und M. Bálint: Über die chromatographische Trennung einiger Enzyme des Emulsins. Enzymologia **5**, 302–306 (1938).
25. L. Zechmeister, G. Tóth und P. Fürth (mitbearbeitet von J. Bársony): Über die chromatographische Trennbarkeit einiger β-Glucosidasen. Enzymologia **9**, 155–160 (1940).
26. L. Zechmeister, W. Grassmann, G. Tóth und R. Bender: Über die Verknüpfungsart der Glucosamin-Reste im Chitin. Ber. dtsch. chem. Ges. **65**, 1706–1708 (1932).
Mat. termtud. ért. **49**, 190–195 (1932).

27. W. GRASSMANN, L. ZECHMEISTER, R. BENDER und G. TÓTH: Über die Chitin-Spaltung durch Emulsin-Präparate (III. Mitt. über enzymatische Spaltung von Polysacchariden). Ber. dtsch. chem. Ges. **67**, 1–5 (1934).
28. L. ZECHMEISTER und G. TÓTH: Chromatographische Zerlegung der Chitinase. Naturwiss. **27**, 367 (1939).
29. L. ZECHMEISTER und G. TÓTH: Chromatographie der in der Chitinreihe wirksamen Enzyme des Emulsins. Enzymologia **7**, 165–169 (1939).
30. L. ZECHMEISTER, G. TÓTH und É. VAJDA: Chromatographie der in der Chitinreihe wirksamen Enzyme der Weinbergschnecke *(Helix pomatia)*. Enzymologia **7**, 170–175 (1939).
31. L. ZECHMEISTER and M. ROHDEWALD: The Detection of Enzymes by the Chromatographic Brush Method. I. Enzymologia **13**, 388–392 (1949).
32. M. ROHDEWALD and L. ZECHMEISTER: The Detection of Enzymes by the Chromatographic Brush Method. II. Enzymologia **15**, 109–114 (1951).
33. L. ZECHMEISTER and M. ROHDEWALD: Some Aspects of Enzyme Chromatography. Fortschr. Chem. organ. Naturstoffe **8**, 341–364 (1951).

Anthocyanes.

34. R. WILLSTÄTTER und L. ZECHMEISTER: Synthese des Pelargonidins. Sitzungsber. kgl. preuß. Akad. Wiss. **34**, 886–893 (1914). Magy. chem. foly. **25**, 1–8 (1919).
35. R. WILLSTÄTTER, L. ZECHMEISTER und W. KINDLER: Synthese des Pelargonidins und Cyanidins. Ber. dtsch. chem. Ges. **57**, 1938–1944 (1924).

Carotenoids in General.

36. L. ZECHMEISTER: Carotinoide höherer Pflanzen (Polyen-Farbstoffe). In: G. KLEIN, Handbuch der Pflanzenanalyse, Bd. III, SS. 1239–1350. Wien: Julius Springer. 1932.
37. L. ZECHMEISTER: Carotinoide. Ein biochemischer Bericht über pflanzliche und tierische Polyenfarbstoffe. XII, 338 SS., mit 85 Abb. Berlin: Julius Springer. 1934.
38. L. ZECHMEISTER: Les caroténoïdes, leurs rapports avec d'autres composés naturels et leur importance biologique. Bull. Soc. chim. biol. (Paris) **16**, 993–1008 (1934).
39. L. ZECHMEISTER: Die Carotinoide und ihre physiologische Bedeutung. Chem.-Ztg. **60**, 505—508 (1936).
40. L. ZECHMEISTER: Die Forschungen Richard Willstätters auf dem Gebiete der Carotinoide. Naturwiss. **20**, 608–612 (1932).

41. L. Zechmeister: Nomenclature of Carotenoid Pigments. Chem. Eng. News **25**, 463–464 (1947).

42. F. J. Petracek and L. Zechmeister: Determination of Partition Coefficients of Carotenoids as a Tool in Pigment Analysis. Analyt. Chemistry **28**, 1484–1485 (1956).

Plant Carotenoids.

43. L. Zechmeister, L. v. Cholnoky und V. Vrabély: Über die katalytische Hydrierung von Carotin. Ber. dtsch. chem. Ges. **61**, 566–568 (1928).
Magy. chem. foly. **34**, 185–193 (1928) [with P. Tuzson].

44. L. Zechmeister und L. v. Cholnoky: Beitrag zum Konstitutions-Problem des Carotins. Ber. dtsch. chem. Ges. **61**, 1534–1539 (1928).

45. L. Zechmeister und V. Vrabély: Zur Deutung der colorimetrischen Hydrierungskurve von Carotinoiden. Ber. dtsch. chem. Ges. **62**, 2232–2235 (1929).

46. L. Zechmeister, L. v. Cholnoky und V. Vrabély: Zur Bestimmung der Doppelbindungen im Carotin-Molekül. Ber. dtsch. chem. Ges. **66**, 123–124 (1933).

47. L. Zechmeister and A. Polgár: Silk Oak Flowers as a Source of β-Carotene. J. Biol. Chem. **140**, 1–3 (1941).

48. L. Zechmeister and W. A. Schroeder: The Pigment of *Mimulus longiflorus* and the Isolation of its γ-Carotene Component. Arch. Biochemistry **1**, 231–238 (1942).

49. L. Zechmeister and L. Cholnoky: Carotenoids of Hungarian Wheat Flour. J. Biol. Chem. **135**, 31–36 (1940).
Mat. termtud. ért. **59**, 146–154 (1940) [with E. Neumann].

50. L. Zechmeister and R. B. Escue: The Provitamin A Content of American Whole Wheat Flour and Whole Wheat Bread. Proc. Nat. Acad. Sci. (U. S. A.) **27**, 528–532 (1941).

51. A. Sandoval and L. Zechmeister: Lycopene. Biochemical Preparations **1**, 57–63 (1949).

52. L. Zechmeister und L. v. Cholnoky: Über das Pigment der reifen Beeren des *Tamus communis*. Ber. dtsch. chem. Ges. **63**, 422–427 (1930).

53. L. Zechmeister und L. v. Cholnoky: Lycopin aus *Solanum dulcamara*. Ber. dtsch. chem. Ges. **63**, 787–790 (1930).
Mat. termtud. ért. **47**, 209–218 (1930).

54. L. Zechmeister und P. Tuzson: Der Farbstoff der Wasser-Melone. Ber. dtsch. chem. Ges. **63**, 2881–2883 (1930).
Magy. chem. foly. **36**, 180–184 (1930).

55. L. Zechmeister and A. Polgár: The Carotenoid and Provitamin A Content of the Watermelon. J. Biol. Chem. **139**, 193–198 (1941).

56. L. ZECHMEISTER und L. v. CHOLNOKY: Über den Farbstoff der Ringelblume *(Calendula officinalis)*. Ein Beitrag zur Kenntnis des Blüten-Lycopins. Z. physiol. Chem. (Hoppe-Seyler) **208**, 26–32 (1932). Mat. termtud. ért. **49**, 181–189 (1932).

57. L. ZECHMEISTER und L. v. CHOLNOKY: Über einen neuen Farbstoff mit Lycopin-Spektrum. Naturwiss. **23**, 407 (1935).

58. L. ZECHMEISTER und L. v. CHOLNOKY: Lycoxanthin und Lycophyll, zwei natürliche Derivate des Lycopins. Ber. dtsch. chem. Ges. **69**, 422–429 (1936).

59. L. ZECHMEISTER und P. TUZSON: Zur Kenntnis des Xanthophylls. I. Katalytische Hydrierung. Ber. dtsch. chem. Ges. **61**, 2003–2009 (1928).

60. L. ZECHMEISTER und P. TUZSON: Zur Kenntnis des Xanthophylls (II. Mitt.). Ber. dtsch. chem. Ges. **62**, 2226–2232 (1929).

61. L. ZECHMEISTER und P. TUZSON: Über das Pigment der Orangenschale. Naturwiss. **19**, 307 (1931).

62. L. ZECHMEISTER und P. TUZSON: Über das Polyen-Pigment der Orange, I. Mitt. Ber. dtsch. chem. Ges. **69**, 1878–1884 (1936).

63. L. ZECHMEISTER und P. TUZSON: Über das Polyen-Pigment der Orange, II. Mitt.: Citraurin. Ber. dtsch. chem. Ges. **70**, 1966–1969 (1937).

64. L. ZECHMEISTER und P. TUZSON: Zur Kenntnis des Mandarinenpigments. I. Z. physiol. Chem. (Hoppe-Seyler) **221**, 278–280 (1933). Mat. termtud. ért. **51**, 266–270 (1934).

65. L. ZECHMEISTER und P. TUZSON: Zur Kenntnis des Mandarinenpigments. II. Z. physiol. Chem. (Hoppe-Seyler) **240**, 191–194 (1936).

66. L. ZECHMEISTER und P. TUZSON: Das Pigment der *Cucurbita maxima* Duch. (Riesenkürbis). Ber. dtsch. chem. Ges. **67**, 824–828 (1934).

67. L. ZECHMEISTER und P. TUZSON: Über den Farbstoff der Sonnenblume (ein Beitrag zur Kenntnis der Blüten-Xanthophylle). Ber. dtsch. chem. Ges. **63**, 3203–3207 (1930).

68. L. ZECHMEISTER und P. TUZSON: Über den Farbstoff der Sonnenblume (II. Mitt.). Ber. dtsch. chem. Ges. **67**, 170–173 (1934). Mat. termtud. ért. **52**, 80–85 (1934).

69. L. ZECHMEISTER, T. BÉRES und E. UJHELYI: Zur Pigmentierung der reifenden Kürbis-Blüte *(Cucurbita pepo)*. Ber. dtsch. chem. Ges. **68**, 1321–1323 (1935).

70. L. ZECHMEISTER, T. BÉRES und E. UJHELYI: Zur Pigmentierung der reifenden Kürbis-Blüte (II. Mitt.). Ber. dtsch. chem. Ges. **69**, 573–574 (1936).

71. A. L. LEROSEN and L. ZECHMEISTER: The Carotenoid Pigments of the Fruit of *Celastrus scandens* L. Arch. Biochemistry **1**, 17–26 (1942).

72. J. H. PINCKARD, J. S. KITTREDGE, D. L. FOX, F. T. HAXO and L. ZECHMEISTER: Pigments from a Marine "Red Water" Population of the Dinoflagellate *Prorocentrum micans*. Arch. Biochem. Biophys. **44**, 189–199 (1953).

73. F. J. PETRACEK and L. ZECHMEISTER: The Structure of Canthaxanthin. Arch. Biochem. Biophys. **61**, 137–139 (1956).

74. L. ZECHMEISTER und L. v. CHOLNOKY: Untersuchungen über den Paprika-Farbstoff. I. Liebigs Ann. Chem. **454**, 54–71 (1927). Mat. termtud. ért. **44**, 404–419 (1927).

75. L. ZECHMEISTER und L. v. CHOLNOKY: Untersuchungen über den Paprika-Farbstoff. II. Liebigs Ann. Chem. **455**, 70–81 (1927). Magy. chem. foly. **32**, 97–102 (1926).

76. L. ZECHMEISTER und L. v. CHOLNOKY: Untersuchungen über den Paprika-Farbstoff. III. (Katalytische Hydrierung.) Liebigs Ann. Chem. **465**, 288–299 (1928).
Mat. termtud. ért. **45**, 639–651 (1928) [with V. VRABÉLY].

77. L. ZECHMEISTER und L. v. CHOLNOKY: Untersuchungen über den Paprika-Farbstoff. IV. Einige Umwandlungen des Capsanthins. Liebigs Ann. Chem. **478**, 95–111 (1930).

78. L. ZECHMEISTER und L. v. CHOLNOKY: Untersuchungen über den Paprika-Farbstoff. V. Natürliche und synthetische Ester des Capsanthins. Liebigs Ann. Chem. **487**, 197–213 (1931).

79. L. ZECHMEISTER und L. v. CHOLNOKY: Untersuchungen über den Paprika-Farbstoff. VI. Das Pigment des japanischen Paprikas. Liebigs Ann. Chem. **489**, 1–6 (1931).

80. L. ZECHMEISTER und L. v. CHOLNOKY: Untersuchungen über den Paprika-Farbstoff. VII. Adsorptionsanalyse des Pigments. Liebigs Ann. Chem. **509**, 269–287 (1934).

81. L. ZECHMEISTER und L. v. CHOLNOKY: Untersuchungen über den Paprika-Farbstoff. VIII. Zur Konstitution des Capsanthins und Capsorubins. Liebigs Ann. Chem. **516**, 30–45 (1935).

82. L. ZECHMEISTER und L. v. CHOLNOKY: Untersuchungen über den Paprika-Farbstoff. IX. Partieller Abbau des Capsanthins. Liebigs Ann. Chem. **523**, 101–118 (1936).

83. L. ZECHMEISTER und L. v. CHOLNOKY: Untersuchungen über den Paprika-Farbstoff. X. Citraurin aus Capsanthin. Liebigs Ann. Chem. **530**, 291–300 (1937).

84. L. ZECHMEISTER und L. v. CHOLNOKY: Über den Zustand der sauerstoffhaltigen Carotinoide in der Pflanze. Vorl. Mitt. Z. physiol. Chem. (Hoppe-Seyler) **189**, 159–161 (1930).

85. L. ZECHMEISTER und L. v. CHOLNOKY: Über den Farbstoff der Bocksdorn-Beere und über das Vorkommen von chemisch gebundenen Carotinoiden in der Natur. Liebigs Ann. Chem. **481**, 42–56 (1930).
Magy. chem. foly. **38**, 85–95 (1932).

86. L. ZECHMEISTER und K. SZILÁRD: Über ein Carotinoid aus den Samenhüllen des Spindelbaumes *(Evonymus europaeus)*. Z. physiol. Chem. (Hoppe-Seyler) **190**, 67–71 (1930).

87. L. ZECHMEISTER und P. TUZSON: Über das Carotinoid des Spindelbaumes *(Evonymus europaeus)*. II. Mitt. Z. physiol. Chem. (Hoppe-Seyler) **196**, 199–200 (1931).

Connected paper: *131*.

Animal Carotenoids.

88. L. ZECHMEISTER und P. TUZSON: Zur Kenntnis des Lipochroms höherer Tiere und des Menschen. Naturwiss. **23**, 680–685 (1935).

89. L. ZECHMEISTER: Lipochrom und Vitamin A. In: H. SCHÖNFELD und G. HEFTER, Chemie und Technologie der Fette und Fettprodukte, Bd. I, SS. 149–193. Wien: Julius Springer. 1936.

90. L. ZECHMEISTER und P. TUZSON: Zur Kenntnis der tierischen Fettfarbstoffe. Ber. dtsch. chem. Ges. **67**, 154–155 (1934).

91. L. ZECHMEISTER: Die Carotinoide im tierischen Stoffwechsel. Ergebn. Physiol. **39**, 117–191 (1937).

92. L. ZECHMEISTER und P. TUZSON: Isolierung des Lipochroms aus Hühner- und Pferdefett. Einige Beobachtungen an menschlichem Fett. Z. physiol. Chem. (Hoppe-Seyler) **225**, 189–195 (1934).

93. L. ZECHMEISTER und P. TUZSON: Isolierung von Komponenten des menschlichen Lipochroms. Z. physiol. Chem. (Hoppe-Seyler) **231**, 259–264 (1935).
Orvosi hetilap **79**, 115–118 (1935).

94. L. ZECHMEISTER et P. TUZSON: Contribution biochimique à l'étude des pigments de la graisse humaine. Bull. Soc. chim. biol. (Paris) **17**, 1110–1118 (1935).

95. L. ZECHMEISTER und P. TUZSON: Über den Lipochrominhalt der menschlichen Leber. Z. physiol. Chem. (Hoppe-Seyler) **234**, 241–244 (1935).

96. L. ZECHMEISTER, P. TUZSON and E. ERNST: Selective Accumulation of Lipochrome. Nature (London) **135**, 1039 (1935).

97. L. ZECHMEISTER und P. TUZSON: Beitrag zum Lipochrom-Stoffwechsel des Pferdes. Z. physiol. Chem. (Hoppe-Seyler) **226**, 255–257 (1934).

98. L. ZECHMEISTER und P. TUZSON: Zur Kenntnis der selektiven Aufnahme von Carotinoiden im Tierkörper. (Zugleich II. Mitt. über den Lipochromstoffwechsel des Pferdes.) Z. physiol. Chem. (Hoppe-Seyler) **234**, 235–240 (1935).

99. L. ZECHMEISTER und P. TUZSON: Notiz über das Lipochrom der Schweineleber. Z. physiol. Chem. (Hoppe-Seyler) **239**, 147–148 (1936).

100. L. ZECHMEISTER und P. TUZSON: Über das Lipochrom des Wasserfrosches *(Rana esculenta)*. Z. physiol. Chem. (Hoppe-Seyler) **238**, 197–203 (1936).

101. B. v. ISSEKUTZ und L. ZECHMEISTER: Notiz über die physiologische Indifferenz des Capsanthins. Biochem. Z. **185**, 1–2 (1927).

Carotenoids: Structural Conversions in vitro.

102. A. POLGÁR and L. ZECHMEISTER: Action of Cold Concentrated Hydriodic Acid on Carotenes: Structure and *cis-trans* Isomerization of Some Reaction Products. J. Amer. Chem. Soc. **65**, 1528–1534 (1943).

103. L. ZECHMEISTER and J. W. SEASE: Conversion of Lutein in a Boric Acid-Naphthalene Melt. I. J. Amer. Chem. Soc. **65**, 1951–1955 (1943).

104. L. ZECHMEISTER and F. J. PETRACEK: On the Structure of the Deoxyluteins. Arch. Biochem. Biophys. **61**, 243–244 (1956).

105. L. WALLCAVE, J. LEEMANN and L. ZECHMEISTER: Action of Boron Trifluoride Etherate on β-Carotene. Proc. Nat. Acad. Sci. (U. S. A.) **39**, 604–606 (1953).

106. L. WALLCAVE and L. ZECHMEISTER: Conversion of Dehydro-β-carotene, *via* its Boron Trifluoride Complex, into an Isomer of Cryptoxanthin. J. Amer. Chem. Soc. **75**, 4495–4498 (1953).

107. F. J. PETRACEK and L. ZECHMEISTER: The Hydrolytic Cleavage Products of Boron Trifluoride Complexes of β-Carotene, Some Dehydrogenated Carotenes, and Anhydrovitamin A_1. J. Amer. Chem. Soc. **78**, 3188–3191 (1956).

108. W. V. BUSH and L. ZECHMEISTER: On Some Cleavage Products of the Boron Trifluoride Complexes of α-Carotene, Lycopene and γ-Carotene. J. Amer. Chem. Soc. **80**, 2991–2999 (1958).

109. L. ZECHMEISTER and L. WALLCAVE: Action of N-Bromosuccinimide on β-Carotene. J. Amer. Chem. Soc. **75**, 4493–4495 (1953).

110. G. KARMAKAR and L. ZECHMEISTER: On Some Dehydrogenation Products of α-Carotene, β-Carotene and Cryptoxanthin. J. Amer. Chem. Soc. **77**, 55–60 (1955).

111. L. ZECHMEISTER and F. J. PETRACEK: Bisdehydro-carotenes. J. Amer. Chem. Soc. **77**, 2567 (1955).

112. F. J. PETRACEK and L. ZECHMEISTER: Reaction of β-Carotene with N-Bromosuccinimide: The Formation and Conversions of Some Polyene Ketones. J. Amer. Chem. Soc. **78**, 1427–1434 (1956).

113. L. ZECHMEISTER: Some in vitro Conversions of Naturally Occurring Carotenoids. Fortschr. Chem. organ. Naturstoffe **15**, 31–82 (1958).

Connected papers: *43, 59, 76, 83, 177, 178.*

Carotenoids: Stereochemistry.

114. L. ZECHMEISTER: *cis-trans* Isomerization and Stereochemistry of Carotenoids and Diphenylpolyenes. Chem. Revs. **34**, 267–344 (1944).

115. L. ZECHMEISTER: Some Stereochemical Aspects of Polyenes. Experientia **10**, 1–11 (1954).

116. L. ZECHMEISTER: Stereoisomeric Provitamins A. Vitamins and Hormones **7**, 57–81 (1949).

117. L. ZECHMEISTER: Les provitamines A stéréoïsomériques. Bull. Soc. chim. biol. (Paris) **31**, 956–964 (1949).

118. K. LUNDE and L. ZECHMEISTER: Infrared Spectra and *cis-trans* Configurations of Some Carotenoid Pigments. J. Amer. Chem. Soc. **77**, 1647–1653 (1955).

119. L. ZECHMEISTER and P. TUZSON: Spontaneous Isomerization of Lycopene. Nature (London) **141**, 249–250 (1938).

120. L. ZECHMEISTER and P. TUZSON: Isomerization of Carotenoids. Biochemic. J. **32**, 1305–1311 (1938).

121. L. ZECHMEISTER und P. TUZSON: Umkehrbare Isomerisierung von Carotinoiden durch Jod-Katalyse. Ber. dtsch. chem. Ges. **72**, 1340–1346 (1939).

122. L. ZECHMEISTER and A. POLGÁR: *cis-trans* Isomerization and Spectral Characteristics of Carotenoids and Some Related Compounds. J. Amer. Chem. Soc. **65**, 1522–1528 (1943).

123. L. ZECHMEISTER and A. POLGÁR: *cis-trans* Isomerization and *cis*-Peak Effect in the α-Carotene Set and in Some Other Stereoisomeric Sets. J. Amer. Chem. Soc. **66**, 137–144 (1944).

124. A. POLGÁR and L. ZECHMEISTER: Isomerization of β-Carotene. Isolation of a Stereoisomer with Increased Adsorption Affinity. J. Amer. Chem. Soc. **64**, 1856–1861 (1942).

125. L. ZECHMEISTER and A. POLGÁR: Contributions to the Stereochemistry of γ-Carotene. J. Amer. Chem. Soc. **67**, 108–112 (1945).

126. L. ZECHMEISTER and L. WALLCAVE: A Study of Some *cis-trans* Isomeric Dehydro-β-carotenes. J. Amer. Chem. Soc. **75**, 5341–5344 (1953).

127. L. ZECHMEISTER, L. v. CHOLNOKY und A. POLGÁR: Über die Isomerisierung des Zeaxanthins und Physaliens. Ber. dtsch. chem. Ges. **72**, 1678–1685 (1939).

128. L. ZECHMEISTER, L. v. CHOLNOKY und A. POLGÁR: Zur Isomerisierung von Xanthophyllen (Nachtrag). Ber. dtsch. chem. Ges. **72**, 2039–2040 (1939).

129. L. ZECHMEISTER and R. M. LEMMON: Contribution to the Stereochemistry of Cryptoxanthin and Zeaxanthin. J. Amer. Chem. Soc. **66**, 317–322 (1944).

130. A. CHATTERJEE and L. ZECHMEISTER: On Some Stereoisomeric Cryptoxanthins. J. Amer. Chem. Soc. **72**, 254–256 (1950).

131. L. ZECHMEISTER und L. v. CHOLNOKY: Untersuchungen über den Paprika-Farbstoff. XI. Isomerisierungs-Erscheinungen. Liebigs Ann. Chem. **543**, 248–257 (1940).

132. A. POLGÁR and L. ZECHMEISTER: A Spectroscopic Study in the Stereoisomeric Capsanthin Set. *cis*-Peak Effect and Configuration. J. Amer. Chem. Soc. **66**, 186–190 (1944).

133. L. ZECHMEISTER and W. A. SCHROEDER: *cis-trans* Isomerization and Spectral Characteristics of Gazaniaxanthin. Further Evidence of its Structure. J. Amer. Chem. Soc. **65**, 1535–1540 (1943).

134. A. POLGÁR, C. B. VAN NIEL and L. ZECHMEISTER: Studies on the Pigments of the Purple Bacteria. II. A Spectroscopic and Stereochemical Investigation of Spirilloxanthin. Arch. Biochemistry **5**, 243–264 (1944).

135. CH. GANSSER und L. ZECHMEISTER: Über einige *cis*-Formen des Canthaxanthins. Helv. Chim. Acta **40**, 1757–1767 (1957).

136. K. TSUKIDA and L. ZECHMEISTER: The Stereoisomerization of β-Carotene Epoxides and the Simultaneous Formation of Furanoid Oxides. Arch. Biochem. Biophys. **74**, 408–426 (1958).

137. L. ZECHMEISTER and E. F. MAGOON: Spectral Maxima of Stereoisomeric Polyenes. Chem. and Ind. **1957**, 431.

Connected papers: *138–152, 174, 175.*

Naturally Occurring *cis* and Poly*cis* Carotenoids.

138. L. ZECHMEISTER and W. A. SCHROEDER: On the Occurrence of Stereoisomeric Carotenoids in Nature. Science (Washington) **94**, 609–610 (1941).

139. L. ZECHMEISTER, A. L. LEROSEN, F. W. WENT and L. PAULING: Prolycopene, a Naturally Occurring Stereoisomer of Lycopene. Proc. Nat. Acad. Sci. (U. S. A.) **27**, 468–474 (1941).

140. A. L. LEROSEN and L. ZECHMEISTER: Prolycopene. J. Amer. Chem. Soc. **64**, 1075–1079 (1942).

141. L. ZECHMEISTER and W. A. SCHROEDER: Pro-γ-carotene. J. Amer. Chem. Soc. **64**, 1173–1177 (1942).

142. L. ZECHMEISTER and W. A. SCHROEDER: The Fruit of *Pyracantha angustifolia*: a Practical Source of Pro-γ-carotene and Prolycopene. J. Biol. Chem. **144**, 315–320 (1942).

143. L. ZECHMEISTER and J. H. PINCKARD: Some Poly*cis*-lycopenes Occurring in the Fruit of *Pyracantha*. J. Amer. Chem. Soc. **69**, 1930–1935 (1947).

144. L. ZECHMEISTER and R. B. ESCUE: Isolation of Prolycopene and Pro-γ-carotene from *Evonymus fortunei*. J. Biol. Chem. **144**, 321–323 (1942).

145. A. J. HAAGEN-SMIT, J. H. PINCKARD and L. ZECHMEISTER: Contribution to the Structure of Pro-γ-carotene and Prolycopene Obtained from Various Sources. Arch. Biochemistry **26**, 358–360 (1950).

146. L. ZECHMEISTER, A. L. LEROSEN, W. A. SCHROEDER, A. POLGÁR and L. PAULING: Spectral Characteristics and Configuration of Some Stereoisomeric Carotenoids Including Prolycopene and Pro-γ-carotene. J. Amer. Chem. Soc. **65**, 1940–1951 (1943).

147. L. ZECHMEISTER and F. J. PETRACEK: Absence of Detectable Poly-*cis* Forms from Heat-Isomerized Lycopene Solutions. J. Amer. Chem. Soc. **74**, 282 (1952).

148. H. J. DEUEL, Jr., J. GANGULY, B. K. KOE and L. ZECHMEISTER: Stereoisomerization of the Poly*cis* Compounds, Pro-γ-carotene and Prolycopene in Chickens and Hens. Arch. Biochem. Biophys. **33**, 143–149 (1951).

149. E. F. MAGOON and L. ZECHMEISTER: Stepwise Stereoisomerization of Prolycopene, a Poly*cis* Carotenoid, to all-*trans*-Lycopene. Arch. Biochem. Biophys. **69**, 535–547 (1957).

150. E. F. MAGOON and L. ZECHMEISTER: On a *cis*-Neurosporene *ex Pyracantha* and the *in vitro* Stereoisomerization of Neurosporene. Arch. Biochem. Biophys. **68**, 263–269 (1957).

151. L. ZECHMEISTER and R. B. ESCUE: New Stereoisomers of Methylbixin. Science (Washington) **96**, 229–230 (1942).

152. L. ZECHMEISTER and R. B. ESCUE: A Stereochemical Study of Methylbixin. J. Amer. Chem. Soc. **66**, 322–330 (1944).

Provitamin A Potency and Structure.

153. H. J. DEUEL, Jr., E. R. MESERVE, C. H. JOHNSTON, A. POLGÁR and L. ZECHMEISTER: Re-investigation of the Relative Provitamin A Potencies of Cryptoxanthin and β-Carotene. Arch. Biochemistry **7**, 447–450 (1945).

154. S. M. Greenberg, A. Chatterjee, C. E. Calbert, H. J. Deuel, Jr. and L. Zechmeister: A Comparison of the Provitamin A Activity of β-Carotene and Cryptoxanthin in the Chick. Arch. Biochemistry 25, 61–65 (1950).

155. H. J. Deuel, Jr., J. Ganguly, L. Wallcave and L. Zechmeister: Provitamin A Activity of a Structural Isomer of Cryptoxanthin and its Methyl Ether. Arch. Biochem. Biophys. 47, 237–240 (1953).

Provitamin A Potency and Steric Configuration.

156. H. J. Deuel, Jr., C. H. Johnston, E. Sumner, A. Polgár and L. Zechmeister: Stereochemical Configuration and Provitamin A Activity. I. All-*trans*-β-carotene and Neo-β-carotene U. Arch. Biochemistry 5, 107–114 (1944).

157. H. J. Deuel, Jr., C. H. Johnston, E. Sumner, A. Polgár, W. A. Schroeder and L. Zechmeister: Stereochemical Configuration and Provitamin A Activity. II. All-*trans*-γ-carotene and Pro-γ-carotene. Arch. Biochemistry 5, 365–371 (1944).

158. H. J. Deuel, Jr., E. Sumner, C. H. Johnston, A. Polgár and L. Zechmeister: Stereochemical Configuration and Provitamin A Activity. III. All-*trans*-α-carotene and Neo-α-carotene U. Arch. Biochemistry 6, 157–161 (1945).

159. H. J. Deuel, Jr., C. H. Johnston, E. R. Meserve, A. Polgár and L. Zechmeister: Stereochemical Configuration and Provitamin A Activity. IV. Neo-α-carotene B and Neo-β-carotene B. Arch. Biochemistry 7, 247–255 (1945).

160. H. J. Deuel, Jr., E. R. Meserve, A. Sandoval and L. Zechmeister: Stereochemical Configuration and Provitamin A Activity. V. Neocryptoxanthin A. Arch. Biochemistry 10, 491–496 (1946).

161. H. J. Deuel, Jr., C. Hendrick, E. Straub, A. Sandoval, J. H. Pinckard and L. Zechmeister: Stereochemical Configuration and Provitamin A Activity. VI. Some *cis-trans* Isomers of γ-Carotene. Arch. Biochemistry 14, 97–103 (1947).

162. H. J. Deuel, Jr., S. M. Greenberg, E. Straub, T. Fukui, A. Chatterjee and L. Zechmeister: Stereochemical Configuration and Provitamin A Activity. VII. Neocryptoxanthin U. Arch. Biochemistry 23, 239–241 (1949).

163. L. Zechmeister, J. H. Pinckard, S. M. Greenberg, E. Straub, T. Fukui and H. J. Deuel, Jr.: Stereochemical Configuration and Provitamin A Activity. VIII. Pro-γ-carotene (a Poly-*cis* Compound) and its All-*trans* Isomer in the Rat. Arch. Biochemistry 23, 242–245 (1949).

164. S. M. Greenberg, C. E. Calbert, J. H. Pinckard, H. J. Deuel, Jr. and L. Zechmeister: Stereochemical Configuration and Provitamin A

Activity. IX. A Comparison of All-*trans*-γ-carotene and Pro-γ-carotene with All-*trans*-β-carotene in the Chick. Arch. Biochemistry **24**, 31–39 (1949).

165. L. ZECHMEISTER, H. J. DEUEL, Jr., H. H. INHOFFEN, J. LEEMANN, S. M. GREENBERG and J. GANGULY: Stereochemical Configuration and Provitamin A Activity. X. A Comparison of Synthetic 15,15'-Mono*cis*-β-carotene (Central Mono*cis*-β-carotene) with All-*trans*-β-carotene in the Rat and Chick. Arch. Biochem. Biophys. **36**, 80–88 (1952).

166. H. J. DEUEL, Jr., H. H. INHOFFEN, J. GANGULY, L. WALLCAVE and L. ZECHMEISTER: Stereochemical Configuration and Provitamin A Activity. XI. A Comparison of Synthetic All-*trans*- and Di*cis*-16,16'-homo-β-carotene $C_{42}H_{58}$ with All-*trans*-β-carotene in the Rat. Arch. Biochem. Biophys. **40**, 352–357 (1952).

Connected papers: *50, 55.*

Naturally Occurring Colorless Polyenes.

167. L. ZECHMEISTER and A. POLGÁR: On the Occurrence of a Fluorescing Polyene with a Characteristic Spectrum. Science (Washington) **100**, 317–318 (1944).

168. L. ZECHMEISTER and J. H. PINCKARD: On Naturally Occurring Colorless Polyenes. Experientia **4**, 474–475 (1948).

169. L. ZECHMEISTER and A. SANDOVAL: Phytofluene. J. Amer. Chem. Soc. **68**, 197–201 (1946).
Bol. Inst. quím. (Mexico) **2**, 3–17 (1946).

170. L. ZECHMEISTER and F. HAXO: Phytofluene in *Neurospora*. Arch. Biochemistry **11**, 539–541 (1946).

171. L. ZECHMEISTER and A. SANDOVAL: The Coloration Given by Vitamin A and Other Polyenes on Acid Earths. Science (Washington) **101**, 585 (1945).

172. L. ZECHMEISTER and A. SANDOVAL: The Occurrence and Estimation of Phytofluene in Plants. Arch. Biochemistry **8**, 425–430 (1945).
Ciencia e Investigación (Buenos Aires) **3**, 314–318 (1947).

173. L. ZECHMEISTER and G. KARMAKAR: The Occurrence of Phytofluene in Green Plant Organs. Arch. Biochem. Biophys. **47**, 160–164 (1953).

174. F. J. PETRACEK and L. ZECHMEISTER: Stereoisomeric Phytofluenes. J. Amer. Chem. Soc. **74**, 184–186 (1952).

175. B. K. KOE and L. ZECHMEISTER: Preparation and Spectral Characteristics of all-*trans*- and a *cis*-Phytofluene. Arch. Biochem. Biophys. **46**, 100–104 (1953).

176. B. K. KOE and L. ZECHMEISTER: *In Vitro* Conversion of Phytofluene and Phytoene into Carotenoid Pigments. Arch. Biochem. Biophys. **41**, 236–238 (1952).

177. L. ZECHMEISTER and B. K. KOE: Stepwise Dehydrogenation of the Colorless Polyenes Phytoene and Phytofluene with N-Bromosuccinimide to Carotenoid Pigments. J. Amer. Chem. Soc. **76**, 2923–2926 (1954).

178. W. LIJINSKY and L. ZECHMEISTER: On the Existence of Some Intermediate Products in the Catalytic Hydrogenation of Lycopene. Arch. Biochem. Biophys. **52**, 358–361 (1954).

179. A. SANDOVAL, E. R. MESERVE, H. J. DEUEL, Jr. and L. ZECHMEISTER: Behavior of Phytofluene in the Animal Body. Arch. Biochemistry **11**, 373–375 (1946).

Miscellaneous Colorless Plant Products.

180. L. ZECHMEISTER und P. SZÉCSI: Notiz über ein Vorkommen von Fumarsäure und von Inosit. Ber. dtsch. chem. Ges. **54**, 172–173 (1921). Magy. chem. foly. **26**, 69–71 (1920).

181. L. ZECHMEISTER und P. TUZSON: Über das Phytosterin der Brennnessel. Z. physiol. Chem. (Hoppe-Seyler) **183**, 74–77 (1929).

182. L. ZECHMEISTER und P. TUZSON: Über eine sterinartige Verbindung aus den Kelchblättern der Sonnenblume. Z. physiol. Chem. (Hoppe-Seyler) **192**, 22–24 (1930).

183. L. ZECHMEISTER und P. TUZSON: Über einige farblose Begleiter von Pflanzencarotinoiden. Z. physiol. Chem. (Hoppe-Seyler) **238**, 204–209 (1936).

184. L. ZECHMEISTER and J. W. SEASE: A Blue-fluorescing Compound, Terthienyl, Isolated from Marigolds. J. Amer. Chem. Soc. **69**, 273–275 (1947).

Plant Physiology and Chemical Genetics.

185. F. W. WENT, A. L. LEROSEN and L. ZECHMEISTER: Effect of External Factors on Tomato Pigments as Studied by Chromatographic Methods. Plant Physiol. 17, 91–100 (1942).

186. A. L. LEROSEN, F. W. WENT and L. ZECHMEISTER: Relation Between Genes and Carotenoids of the Tomato. Proc. Nat. Acad. Sci. (U. S. A.) **27**, 236–242 (1941).

187. J. BONNER, A. SANDOVAL, Y. W. TANG and L. ZECHMEISTER: Changes in Polyene Synthesis Induced by Mutation in a Red Yeast *(Rhodotorula rubra)*. Arch. Biochemistry **10**, 113–123 (1946).

188. Y. W. TANG, J. BONNER and L. ZECHMEISTER: Some Further Experiments on Red Yeast Polyenes. Arch. Biochemistry **21**, 455–456 (1949).

189. L. ZECHMEISTER and F. W. WENT: Some Stereochemical Aspects in Genetics. Nature (London) **162**, 847 (1948).

cis-trans Isomerization of Diphenylpolyenes, Phenyl-polyenaldehydes, Azines, and Cyanines.

190. L. ZECHMEISTER and W. H. MCNEELY: Separation of *cis* and *trans* Stilbenes by Application of the Chromatographic Brush Method. J. Amer. Chem. Soc. **64**, 1919–1921 (1942).

191. A. SANDOVAL and L. ZECHMEISTER: Some Spectroscopic Changes Connected with the Stereoisomerization of Diphenylbutadiene. J. Amer. Chem. Soc. **69**, 553–557 (1947).

192. J. H. PINCKARD, B. WILLE and L. ZECHMEISTER: A Comparative Study of the Three Stereoisomeric 1,4-Diphenylbutadienes. J. Amer. Chem. Soc. **70**, 1938–1944 (1948).

193. K. LUNDE and L. ZECHMEISTER: *cis-trans* Isomeric 1,6-Diphenylhexatrienes. J. Amer. Chem. Soc. **76**, 2308–2313 (1954).

194. E. F. MAGOON and L. ZECHMEISTER: On the *cis* Forms of Some Biphenylene Derivatives of Butadiene and Hexatriene. J. Amer. Chem. Soc. **77**, 5642–5646 (1955).

195. L. ZECHMEISTER and A. L. LEROSEN: Contribution to the Stereochemistry of Diphenylpolyenes. Science (Washington) **95**, 587–588 (1942).

196. L. ZECHMEISTER and A. L. LEROSEN: Stereoisomeric Diphenyloctatetraenes. J. Amer. Chem. Soc. **64**, 2755–2759 (1942).

197. L. ZECHMEISTER and J. H. PINCKARD: Stereoisomeric Diphenyloctatetraenes. II. J. Amer. Chem. Soc. **76**, 4144–4148 (1954).

198. K. LUNDE and L. ZECHMEISTER: A Study of the Infrared Spectra of Some Stereoisomeric Diphenylpolyenes. Acta Chem. Scand. **8**, 1421–1432 (1954).

199. CH. GANSSER and L. ZECHMEISTER: On Some *cis*-Forms of Phenylundecapentaenal. J. Amer. Chem. Soc. **79**, 3854–3858 (1957).

200. J. DALE and L. ZECHMEISTER: On the Stereochemistry of Azines: Cinnamalazine and Phenylpentadienalazine. J. Amer. Chem. Soc. **75**, 2379–2386 (1953).

201. L. ZECHMEISTER and J. H. PINCKARD: On Stereoisomerism in the Cyanine Dye Series. Experientia **9**, 16–17 (1953).

Carcinogens and Related Compounds.

202. M. B. SHIMKIN, B. K. KOE and L. ZECHMEISTER: An Instance of the Occurrence of Carcinogenic Substances in Certain Barnacles. Science (Washington) **113**, 650–651 (1951).

203. L. ZECHMEISTER and B. K. KOE: The Isolation of Carcinogenic and Other Polycyclic Aromatic Hydrocarbons from Barnacles. Arch. Biochem. Biophys. **35**, 1–11 (1952).

204. B. K. KOE and L. ZECHMEISTER: The Isolation of Carcinogenic and Other Polycyclic Aromatic Hydrocarbons from Barnacles. II. The Goose Barnacle *(Mitella polymerus)*. Arch. Biochem. Biophys. **41**, 396–403 (1952).

205. L. ZECHMEISTER and W. LIJINSKY: Some Neutral Constituents of a Natural Tar Originating from the LaBrea Pits. Arch. Biochem. Biophys. **47**, 391–395 (1953).

206. B. K. KOE, D. L. FOX and L. ZECHMEISTER: The Nature of Some Fluorescing Substances Contained in a Deep-Sea Mud. Arch. Biochemistry **27**, 449–452 (1950).

207. F. J. PETRACEK, D. L. FOX and L. ZECHMEISTER: The Fluorescing Substances of an Intertidal Ocean Mud. Arch. Biochemistry **30**, 466–468 (1951).

208. D. L. FOX, B. K. KOE, F. J. PETRACEK and L. ZECHMEISTER: Some Fluorescent Substances Contained in the Marine "Blood Worm" *(Thoracophelia mucronata)*. Arch. Biochem. Biophys. **40**, 135–142 (1952).

209. W. LIJINSKY and L. ZECHMEISTER: On the Catalytic Hydrogenation of 3,4-Benzpyrene. J. Amer. Chem. Soc. **75**, 5495–5497 (1953).

210. P. KOTIN, H. L. FALK, W. LIJINSKY and L. ZECHMEISTER: Inhibition of the Effect of Some Carcinogens by their Partially Hydrogenated Derivatives. Science (Washington) **123**, 102 (1956).

211. L. ZECHMEISTER: Antivitamins and Anticarcinogens. Triangle (Basel) **3**, 218–223 (1958).

Light Aromatic Hydrocarbons.

212. I. PFEIFER and L. ZECHMEISTER: Adatok könnyü aromás szénhydrogének pyrogenetikus elöállításához. Magy. chem. foly. **25**, 139–145 (1919).
Nordisk Handelsbl. kem. Ind. **4**, 53 (1923).

Organic Lead Compounds.

213. L. ZECHMEISTER und J. CSABAY: Über einige Umwandlungen von Diphenyl-bleihalogeniden. Ber. dtsch. chem. Ges. **60**, 1617–1621 (1927).

Organic Minerals.

214. L. ZECHMEISTER und V. VRABÉLY: Notiz über Ajkait (ein organisches Mineral aus Ungarn). Ber. dtsch. chem. Ges. **59**, 1426–1428 (1926).
Mat. termtud. ért. **43**, 332–341 (1926).

215. L. ZECHMEISTER und V. VRABÉLY: Über Telegdit, ein fossiles Harz aus Siebenbürgen. Zbl. Mineral., Geol., Paläont., Abt. A, 1927, 287–290.

216. L. ZECHMEISTER, G. TÓTH und A. KOCH: Untersuchung eines neuen fossilen Harzes: Kiscellit. Zbl. Mineral., Geol., Paläont., Abt. A, 1934, 60–61.
Mat. termtud. ért. 51, 502–505 (1934).

217. L. ZECHMEISTER and O. FREHDEN: A Chromatographic Study of Lignite. Nature (London) 144, 331 (1939).

218. L. ZECHMEISTER and W. T. STEWART: The Isolation of Some Compounds from North Dakota Lignite. J. Amer. Chem. Soc. 63, 2851 (1941).

Chromatographic Methods.

219. L. ZECHMEISTER: Mikhail Tswett, The Inventor of Chromatography. Isis (U. S. A.) 36, 108–109 (1946).

220. L. ZECHMEISTER und L. v. CHOLNOKY: Dreißig Jahre Chromatographie. Monatsh. Chem. 68, 68–80 (1936).

221. L. ZECHMEISTER: History, Scope, and Methods of Chromatography. Ann. New York Acad. Sci. 49, 145–160 (1948).

222. L. ZECHMEISTER: Early History of Chromatography. Nature (London) 167, 405 (1951).

223. L. ZECHMEISTER und L. v. CHOLNOKY: Die chromatographische Adsorptionsmethode. XI, 231 SS., mit 45 Abb. Wien: Julius Springer. 1937.

224. L. ZECHMEISTER und L. v. CHOLNOKY: Die chromatographische Adsorptionsmethode. Grundlagen. Methodik. Anwendungen. 2. wesentlich erweiterte Aufl. XIII, 354 SS., mit 74 Abb. Wien: Julius Springer. 1938.

225. L. ZECHMEISTER and L. CHOLNOKY: Principles and Practice of Chromatography. Translated from the 2nd German ed. by A. L. Bacharach and F. A. Robinson. With a Foreword by I. M. Heilbron. XVIII, 362 pp., with 71 figures. London: Chapman & Hall; New York: J. Wiley & Sons, Inc. 1941; 2nd impr. 378 pp. 1943.

226. L. ZECHMEISTER: Progress in Chromatography 1938–1947. XVIII, 368 pp., with 23 figures. London: Chapman & Hall; New York: J. Wiley & Sons, Inc. 1950.

227. L. ZECHMEISTER: A chromatographia néhány ujabb alkalmazásáról. Mat. termtud. ért. 61, 46 (1942).

228. L. ZECHMEISTER: Some Biochemical Studies Based on Chromatographic Methods. The Harvey Lectures 47, 221–241 (1951–1952).

229. L. ZECHMEISTER: Adsorption and Some Constitutional and Steric Properties. Discuss. Faraday Soc. 7, 54–57 (1949).

230. L. ZECHMEISTER: Column Chromatography and Some Applications in Stereochemistry. In: Modern Chemistry for Engineers and Scientists, pp. 124–145. New York: McGraw-Hill. 1957.

231. L. ZECHMEISTER, L. de CHOLNOKY et E. UJHELYI: Contribution à la chromatographie des substances incolores. Bull. Soc. chim. biol. (Paris) **18**, 1885–1887 (1936).

232. L. ZECHMEISTER et O. FREHDEN: Quelques applications de la méthode au pinceau dans l'analyse chromatographique. Bull. Soc. chim. biol. (Paris) **22**, 458–460 (1940).

233. L. ZECHMEISTER: Paper Disk Columns in Glass Chromatographic Tubes. Science (Washington) **113**, 35–36 (1951).

234. J. W. SEASE and L. ZECHMEISTER: Chromatographic and Spectral Characteristics of Some Polythienyls. J. Amer. Chem. Soc. **69**, 270–273 (1947).

235. J. H. PINCKARD, A. CHATTERJEE and L. ZECHMEISTER: The Behavior of Anthrone on Some Alumina Columns. J. Amer. Chem. Soc. **74**, 1603–1604 (1952).

236. L. ZECHMEISTER: Über die chromatographische Bestimmung des Provitamins A. Compt. rend. 5e Congr. internat. ind. agric., Schéveningue 1937, 20–21.

237. L. ZECHMEISTER: Stereochemistry and Chromatography. Ann. New York Acad. Sci. **49**, 220–234 (1948).

238. L. ZECHMEISTER: Chromatography and Spectroscopy in Organic Chemistry and Stereochemistry. Amer. Scientist **36**, 505–516 (1948).
Kagaku (Japan) **19**, 498–503 (1949).

239. L. ZECHMEISTER, O. FREHDEN und P. FISCHER JÖRGENSEN: Chromatographische Trennung von *cis*- und *trans*-Azobenzol. Naturwiss. **26**, 495 (1938).

240. L. ZECHMEISTER, W. H. MCNEELY and G. SÓLYOM: Chromatography of *cis* and *trans* Benzoin and Anisoin Oximes with Application of the Brush Method. J. Amer. Chem. Soc. **64**, 1922–1924 (1942).

Connected papers: *24, 25, 28–33.*

Reduction Methods.

241. N. BJERRUM und L. ZECHMEISTER: Beurteilung und Entwässerung des Methylalkohols mit Hilfe von Magnesium. Ber. dtsch. chem. Ges. **56**, 894–899 (1923).

242. N. BJERRUM und L. ZECHMEISTER: Berichtigung zu unserer Arbeit „Beurteilung und Entwässerung des Methylalkohols mit Hilfe von Magnesium". Ber. dtsch. chem. Ges. **56**, 1247 (1923).

243. L. ZECHMEISTER und P. ROM: Über die Anwendung von Magnesium und Methylalkohol als Reduktionsmittel. Liebigs Ann. Chem. **468**, 117–132 (1929).
Mat. termtud. ért. 45, 619–638 (1928).

244. L. ZECHMEISTER und J. TRUKA: Notiz über die Reduktion von Schiffschen Basen. Ber. dtsch. chem. Ges. **63**, 2883–2884 (1930).

245. L. ZECHMEISTER und P. ROM: Über die Reduktion von Nitro- zu Azoxy-Körpern mit Magnesium und Salmiak-Lösung. Ber. dtsch. chem. Ges. 59, 867–874 (1926).

246. L. ZECHMEISTER und L. V. CHOLNOKY: Ein Apparat zur katalytischen Hydrierung im Halbmikromaßstab. Chem.-Ztg. **60**, 655–656 (1936).

247. L. ZECHMEISTER: Über eine kleine Vereinfachung der Rechnung bei Gasanalysen. Chem.-Ztg. **52**, 887 (1928).

Connected papers: *43*, *45*, *76*.

Effect of Ultrasonic Waves on Organic Compounds.

248. L. ZECHMEISTER and L. WALLCAVE: On the Cleavage of Benzene, Thiophene and Furan Rings by Means of Ultrasonic Waves. J. Amer. Chem. Soc. 77, 2853–2855 (1955).

249. L. ZECHMEISTER and E. F. MAGOON: On the Ultrasonic Cleavage of the Pyridine Ring. J. Amer. Chem. Soc. **78**, 2149–2150 (1956).

250. L. ZECHMEISTER and E. F. MAGOON: Some Experiments with Ultrasonic Waves. Festschrift Arthur Stoll, pp. 59–63. Basel: Birkhäuser. 1957.

251. D. L. CURRELL and L. ZECHMEISTER: On the Ultrasonic Cleavage of Some Aromatic and Heterocyclic Rings. J. Amer. Chem. Soc. **80**, 205–208 (1958).

Electrolytic Dissociation.

252. N. BJERRUM, A. UNMACK und L. ZECHMEISTER: Die Dissoziationskonstante von Methylalkohol. Danske Vidensk. Selsk. Math.-fys. Medd. 5, No. 11 (pp. 1–34) (1924).

Radioactive Isotopes.

253. G. V. HEVESY und L. ZECHMEISTER: Über den intermolekularen Platzwechsel gleichartiger Atome. Ber. dtsch. chem. Ges. 53, 410–415 (1920).

254. G. V. HEVESY und L. ZECHMEISTER: Über den Verlauf des Umwandlungsvorganges isomerer Ionen. Z. Elektrochem. **26**, 151–153 (1920).
Magy. chem. foly. **26**, 58–64 (1920).

Textbooks.

255. L. Zechmeister: Organikus Chemia, 2 Vol. Budapest. 1930–1932.
256. C. Faurholt, J. C. Gjaldbaek and L. Zechmeister: Chemiai Gyakorlatok (Practice in Chemistry). Pécs. 1932.
257. L. Zechmeister: Bevezetés a titrimetriába (Introduction to Titrimetry). Pécs. 1925.

Author Index.

Bálint, M. 2.
Bársony, J. 2.
Bender, R. 2, 3.
Béres, T. 5.
Bjerrum, N. 18, 19.
Bonner, J. 14.
Bush, W. V. 8.

Calbert, C. E. 12.
Chatterjee, A. 10, 12, 18.
Cholnoky, L. v. 4, 5, 6, 7, 10, 17, 18, 19.
Csabay, J. 16.
Currell, D. L. 19.

Dale, J. 15.
Deuel, H. J., Jr. 11, 12, 13, 14.

Ernst, E. 7.
Escue, R. B. 4, 11.

Falk, H. L. 16.
Faurholt, C. 20.
Fischer Jörgensen, P. 18.
Fox, D. L. 6, 16.
Frehden, O. 17, 18.
Fukui, T. 12.
Fürth, P. 2.

Ganguly, J. 11, 12, 13.
Gansser, Ch. 10, 15.
Gjaldbaek, J. C. 20.
Grassmann, W. 2, 3.
Greenberg, S. M. 12, 13.

Haagen-Smit, A. J. 11.
Haxo, F. T. 6, 13.
Hendrick, C. 12.
Hevesy, G. v. 19.

Inhoffen, H. H. 13.
Issekutz, B. v. 8.

Johnston, C. H. 11, 12.

Karmakar, G. 8, 13.
Kindler, W. 3.
Kittredge, J. S. 6.
Koch, A. 17.
Koe, B. K. 11, 13, 14, 15, 16.
Kotin, P. 16.

Leemann, J. 8, 13.
Lemmon, R. M. 10.
LeRosen, A. L. 5, 10, 11, 14, 15.
Lijinsky, W. 14, 16.
Lunde, K. 9, 15.

Magoon, E. F. 10, 11, 15, 19.
Mark, H. 1.
McNeely, W. H. 15, 18.
Meserve, E. R. 11, 12, 14.

Neumann, E. 4.
Niel, C. B. van 10.
Pauling, L. 10, 11.
Petracek, F. J. 4, 6, 8, 9, 11, 13, 16.
Pfeifer, I. 16.
Pinckard, J. H. 6, 11, 12, 13, 15, 18.
Pinczési, I. 2.
Polgár, A. 4, 8, 9, 10, 11, 12, 13.

Rohdewald, M. 3.
Rom, P. 19.

Sandoval, A. 4, 12, 13, 14, 15.
Schroeder, W. A. 4, 10, 11, 12.
Sease, J. W. 8, 14, 18.
Shimkin, M. B. 15.

Sólyom, G. 18.
Stadler, R. 2.
Stewart, W. T. 17.
Straub, E. 12.
Sumner, E. 12.
Szécsi, P. 14.
Szilárd, K. 7.

Tang, Y. W. 14.
Tóth, G. 1, 2, 3, 17.
Truka, J. 19.
Tsukida, K. 10.
Tuzson, P. 4, 5, 7, 8, 9, 14.

Ujhelyi, E. 5, 18.
Unmack, A. 19.

Vajda, É. 3.
Vrabély, V. 4, 6, 16, 17.

Wallcave, L. 8, 9, 12, 13, 19.
Went, F. W. 10, 14, 15.
Wille, B. 15.
Willstätter, R. 1, 3.

Manzsche Buchdruckerei, Wien IX.

GPSR Compliance
The European Union's (EU) General Product Safety Regulation (GPSR) is a set of rules that requires consumer products to be safe and our obligations to ensure this.

If you have any concerns about our products, you can contact us on

ProductSafety@springernature.com

In case Publisher is established outside the EU, the EU authorized representative is:

Springer Nature Customer Service Center GmbH
Europaplatz 3
69115 Heidelberg, Germany

www.ingramcontent.com/pod-product-compliance
Ingram Content Group UK Ltd.
Pitfield, Milton Keynes, MK11 3LW, UK
UKHW041849190726
13854UKWH00002B/786

* 9 7 8 3 7 0 9 1 2 0 3 3 0 *